NORTHERN BEE BOOKS

Scout Bottom Farm, Mytholmroyd, West Yorkshire

www.northernbeebooks.co.uk

SPLITS AND VARROA: *AN INTRODUCTION TO SPLITTING HIVES AS PART OF VARROA CONTROL.*
by William Hesbach

ISBN: 978-1-908904-86-7

First published by Northern Bee Books, 2016

Scout Bottom Farm
Mytholmroyd
Hebden Bridge
HX7 5JS (UK)

www.northernbeebooks.co.uk

Tel: +44 1422 882751

SPLITS AND VARROA

AN INTRODUCTION TO SPLITTING HIVES AS PART OF VARROA CONTROL

by

WILLIAM HESBACH

NORTHERN BEE BOOKS

Scout Bottom Farm, Mytholmroyd, West Yorkshire

www.northernbeebooks.co.uk

CONTENTS

INTRODUCTION

Varroa have a long history with honey bees. They were first identified in Java in 1904 as the parasitic species, Varroa jacobsoni, living in balance with the Asian honey bee Apis cerana. In the US, varroa were first discovered on our European honey bee, Apis mellifera, in 1987 in the states of Wisconsin and Florida. The following year varroa were detected in 12 additional states, and today virtually every honey bee colony in the country has some level of infestation. At some point previous to the US discovery, varroa jumped species from the Asian bee. In doing so, varroa evolved into a separate species now called Varroa destructor. This adaptation proved fatal to the European honey bee because, unlike the Asian honeybee, the European honey bee had not evolved into a balanced relationship that would allow both to co-exist. Left untreated, a colony of European honey bees with even a moderate infestation will eventually weaken and die.

In the years following the discovery, we witnessed the destruction of our feral bee population and the slow infestation of every managed bee colony in the US. Today, after decades, and despite our best efforts, varroa persist as the single most demanding management issue of our beekeeping experience.

Widespread use of toxic chemical controls in the early years proved both environmentally hazardous and easily resisted by varroa. Consequently, various varroa management styles have emerged. Some commercial beekeepers still rely on toxic chemicals, other sideline and hobbyist beekeepers use integrated pest management (IPM) combining chemical and natural controls, and a small number of beekeepers are experimenting with treatment free beekeeping. Regardless of style and the debates concerning results, one single fact remains - beekeepers must manage varroa to levels below damaging thresholds or colonies die. To meet those thresholds, some beekeepers, looking for less toxic alternatives, are beginning the judicious use of softer chemicals like organic acids and compounds from thyme and hops. Then, there are growing numbers of beekeepers looking further for even less toxic and more natural alternatives. That leads us to the topic of this booklet. One dependable way to accomplish a measure of control without chemicals is to manipulate the reproductive events in the varroa lifecycle. What follows is an introduction to achieving those manipulations using different techniques of splitting a colony.

CHAPTER 1

HOW SPLITS WORK

The reason that splits work is varroa's dependence on the developing bee larvae. Varroa can only reproduce by entering a cell containing a mature larva ready to be capped (Figure 1). Once the cell is capped, varroa begin to reproduce. They feed on the developing pupa's hemolymph, vectoring viruses and ultimately weakening the emerging bee. This cycle then repeats for as long as there are bee larvae to infest. Since more than one mature varroa will emerge with a single adult bee, eventually the varroa population can overrun the bee population and the colony's viral load will cause extensive disease and ultimate collapse. Because varroa reproduce exponentially, meaning one female can produce more than one offspring, a colony can collapse in a single season.

There are many ways to make splits and almost an equal number of reasons why beekeepers make them. In this booklet I'll discuss splits made with the intent to control varroa, using manipulations to temporarily stop the colony from producing bee larvae and therefore denying varroa the ability to reproduce.

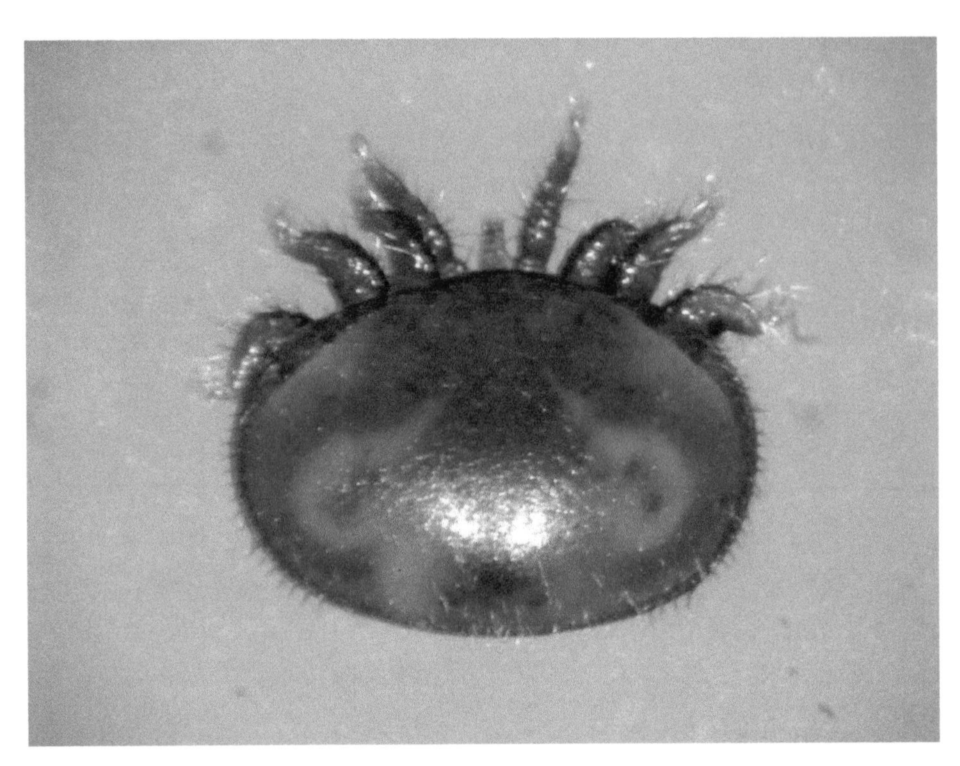

Figure 1: *An adult female varroa has a reddish brown body and is visible to the naked eye. Varroa enter a cell just prior capping and crawl to the bottom under the larva. After crawling under the larva, the female mite will submerge itself in the brood food under the larva where it will remain until the cell is capped. While submerged, the mite erects its peretrimes, which serve as snorkels, allowing the mite to breathe while it is submerged.*

As we proceed through the different techniques, keep in mind that splits are done for the benefit of the entire colony. What this means is that each split you make, plus what's left of the parent colony, must go through a period when no eggs are being laid, either because the splits are in transition to a new queen or the existing queen is being restricted from laying. This results in what is referred to as a brood break, or broodless period, which must be of sufficient duration to deny mites the opportunity to reproduce for an extended period. A sufficient duration means enough time for the emergence of all the capped brood in the colony and especially all the drone brood. This is important because drones remain capped for 3 days longer than worker brood allowing more varroa offspring in drone brood to mature. Consequently, varroa have adapted a strategy that favors drone cells, and studies indicate that varroa are 8 times more likely to choose a drone cell to enter for reproduction.

During the broodless period all the reproducing mites, already under capped cells, will emerge, and with no available larvae for continued reproduction, some will die and others will be groomed off. Since the population is diminished, once the honey bee brood cycle resumes, the emerging bee population can grow faster than the varroa population. This has been described as a forever-young colony, where periodic brood breaks allow bee populations to naturally outrun their pathogens and pests. In nature, the honey bee that most exemplifies a forever-young colony is African honey bee that defeats varroa by constantly swarming and therefore initiating an extended brood break with each newly formed colony.

With a split, accompanied by a brood break, there are a few important considerations. The colony loses a great deal of population, and the split must be timed to allow recovery before the end of the brood rearing season. Also, facilitating a successful brood break requires understanding the brood cycle. The splits must be re-queened in a way where egg laying is delayed. How you introduce a queen will depend on the resources you have, or can obtain, at the time of the split. Here are some options to re-queen and facilitate a brood break:

- A walkaway split where a colony raises an emergency queen.
- Timing the introduction of a ripe queen cell.
- Timing the introduction of a virgin queen.

Let's look at each one with the idea in mind that you have determined a need for varroa control and that splitting the colony is possible. As you read, you will notice each option provides a consistent broodless period, but requires slightly different re-queening timing based on managing when egg laying resumes. I've provided a critical timing chart of different biological events that occur after splits and swarms. It may be helpful to keep them in mind while reading. Also keep in mind, that these numbers are based on the consistent experience of both national queen breeders and local experienced beekeepers. Although they may vary slightly in your situation, they are reliable because they reflect the biology of the bee.

CHAPTER TWO

WALKAWAY SPLITS

The name "walkaway split" describes a procedure where the resources of a robust parent colony are divided, or split, into smaller colonies, and those colonies are basically "walked away from" and allowed to raise new queens. If you have a strong overwintered colony and a high mite count then a walkaway, done early in the season, can be a good choice. This option works best around swarm season with good nectar flow and plenty of available drones. Otherwise any colony needing varroa control can be split as a walkaway as long as it has the resources needed. The main advantage of a walkaway split is that it can be done without the timing restrictions of acquiring new queens or queen cells.

The basic procedure is to divide the resources, making certain that each split has a frame of capped brood and a frame of open brood including eggs. Eggs are important, because they guarantee the correct age larvae for emergency queen cells (Figure 2). Ideally, the splits would also contain a frame with pollen and some honey. In the event you are splitting in less than ideal natural nectar flows,

Figure 2: *This shows nurse bees tending to an emergency queen cell developing in a walkaway spilt. If you look close you can see the queen larva inside this uncapped queen cell bathed in lots of royal jelly. The larva inside may have started from an egg or a one-day-old larva.*

Figure 3: *This frame shows emergency cells that were drawn and capped just 9 days after a queenless split. It illustrates that in an emergency situation the colony will attempt raising many new queens, but only one will ultimately become the queen.*

judge the need and use supplemental feeding. Just note that when splits are done in a dearth, they can be vulnerable so check them for robbing behavior.

About 24 hours after splits, bees will begin to draw queen cells and raise emergency queens (Figure 3). If all goes smoothly, raising a new egg-laying queen will take about 27 days. It will take another 21 days, or about 7 weeks total, before the colony has an emerging population of new bees. So with walkaways, or any split, the time of season is important to consider, and the earlier in the season the better. Ideally, your splits should be queen-right with emerging brood well in advance of your season's winter bee production.

During the first 9-10 days after the split, varroa can still find open cells with mature larvae for reproduction, but after that, all existing larvae will be capped. While the current brood cycle finishes, varroa will continue to emerge with adult bees, but are denied bee larvae for reproduction, until about day 34, depending on when the new queen starts laying. At that point, the new queen's larvae will mature, and varroa can resume reproduction. The 25-day period between day 9, when all brood is capped, and day 34, when larvae are again available for varroa reproduction, is significant for two reasons. First, as mentioned above, it's the period without open brood, when varroa cannot reproduce causing future populations to decline. Second, somewhere around day 24, after the last of the drone brood emerges, the entire varroa population is phoretic (being carried on bees). That window is difficult to predict, but can last about 10 days. That's the sweet spot for optional additional treatments. If you intend to perform additional treatments aimed at killing the phoretic mites, inspections can determine when all the drone brood has emerged or when egg laying has started making new brood available.

Walkaways During Swarm Season

There's a tradition during swarm season where beekeepers who discover a colony with swarm cells, usually in a double deep brood box configuration, simply split a colony in half onto two separate bottom boards. They provision each box with some swarm cells (Figure 4) without regard for the location of the queen. These are sometimes called 50/50 splits, and at times they can work to delay or avert swarms. For varroa control, a 50/50 split will only help the queenless box. Varroa control in the other box will require that queen be restricted from laying. If you are in this situation and want varroa control on the queen-right side, see the parent colony queen section below for information on restricting a laying queen. Also see the notes below on swarm behavior. During swarming, reproductive impulses are strong and the colony is seeking a balance that at times cannot be stopped. It's better to split prior to an emergency swarm situation, because you're better able to manage the colony's reproductive urge.

CHAPTER THREE

RIPE QUEEN CELLS

The goal when using ripe queen cells is to introduce them into a split that you've made about 14 days earlier. You provision the splits the same as a walkaway, with the exception that you do not provision them with early stage larvae or eggs. In other words, unlike a walkaway split, you don't want this split to have the ability to make an emergency queen. So during the 14-day period prior to introduction of your queen cell, it's critical there are no accidental emergency queen cells in production. It's easy to unintentionally put in a frame with a small patch of open brood, so you need to check. Emergency cells are normally started within a day or two, so check a few days after the split, and if you encounter any remove the frame, shake or brush of the bees, destroy all the emergency cells, and put the frame back. Then, before installing the ripe queen cell check for emergency cells again, and repeat the procedure above if needed.

As mentioned, the different options are all seeking the same broodless period. As an example, consider that when a colony is raising an emergency queen, like with a walkaway split, the colony will pick a 1-day-old larva, which is actually 4 days into the 16-day egg to emergency queen cycle. This means that a virgin queen will emerge about 12 days after the walkaway split date. Therefore, by introducing a ripe queen cell into a 14-day-old split, with only 2 days before the queen emerges, you are simply timing the emergence of the virgin queen to match that of a walkaway split – both at 16 days. A few days after the virgin emerges, she begins mating flights, and the events inside the colony follow the timing of a walkaway split, with all the same benefits.

Finally, the advantage of using a ripe queen cell is your choice of queens. If you're doing this during swarm season, you can use your own ripe swarm cells from colonies that have characteristics you want, or you can purchase desired cells from local queen breeders.

How Do You Know When a Queen Cell Is Ripe?

When we talk about a ripe queen cell, we're referring to a cell where, as mentioned, the virgin queen inside is only a few days from emergence (Figure 4). That's commonly considered day 13 or 14 from the time the egg was laid. If you have a reliable source for purchasing ripe cells, you're fortunate that maturity can be discussed with your supplier. Otherwise, if you're trying to determine maturity using your own cells, you'll need to keep track of timing and also learn what a ripe queen cell looks like. Of course, knowing the date the queen cell was capped is the best method. Queen cells are capped on day 8, so in another 5 or 6 days the cell is considered mature. If you don't know the exact timing, there are a few characteristics that help, the most useful being the appearance of the cap and the behavior of the bees around the cell.

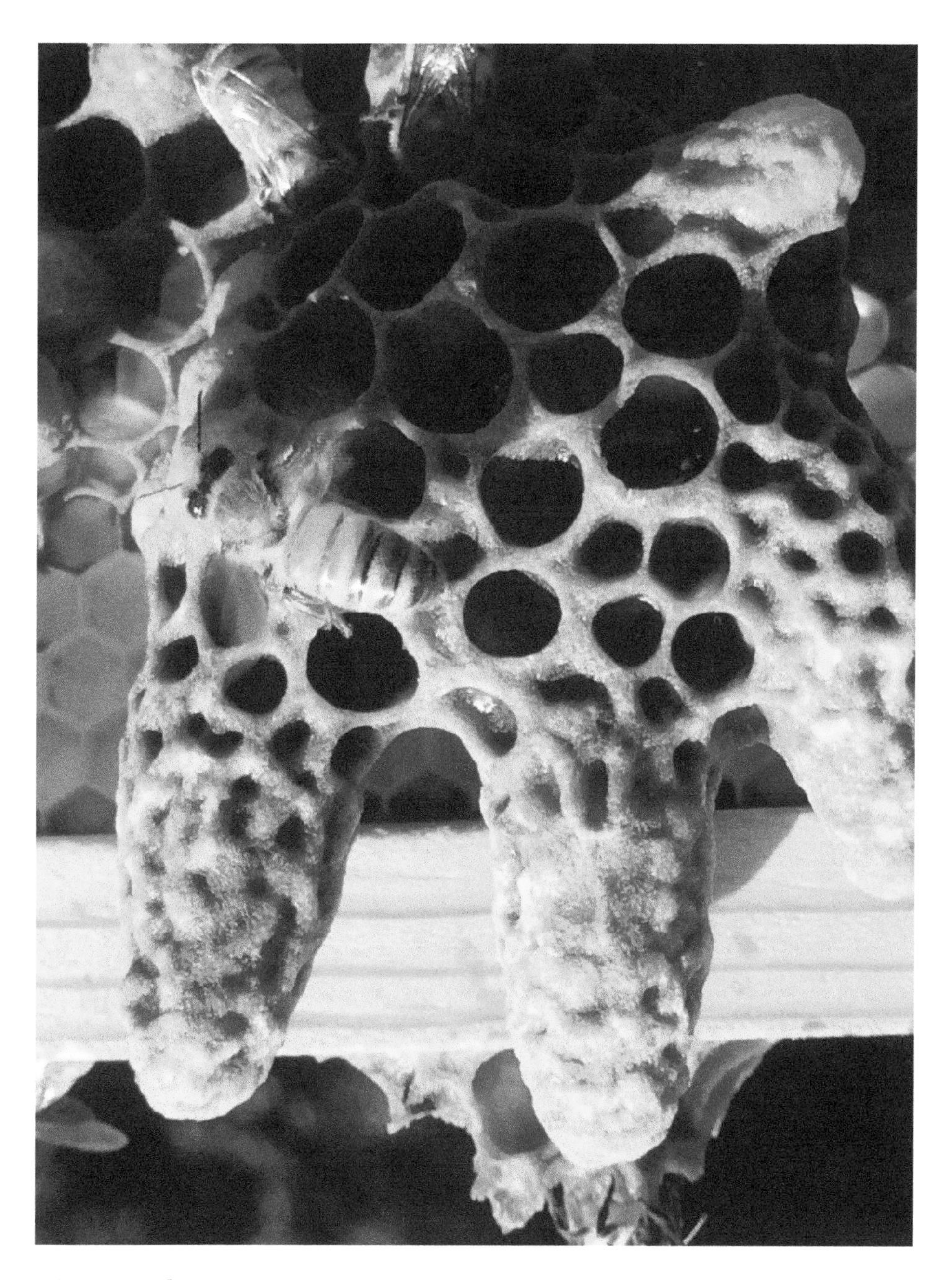

Figure 4: *These are examples of ripe queen cells and in this case they are swarm cells. The best way to tell if a queen cell is ripe is to know when it was capped, or you may be able to purchase one from a reliable source. Also, you can sometimes tell a queen cell is ripe by observing bees crawling on the cell and shaking it to communicate with the queen inside.*

When a queen cell ripens, the workers will begin to chew the wax off the cap, leaving only a thin covering. That changes the appearance in that the very tip of the cap appears to have a different texture than the rest of the cell, and the color is lighter (see photo). When the queen is hours or even moments away from emerging, you can sometimes see a small slit around the base of the cap where she's beginning to chew her way out. If you have a cell with signs that the queen is moments from emerging, you can gently insert the capped end into a queen cage and allow the queen to emerge into the cage. After she has emerged, you can then remove the cell and insert a candy plug. With the virgin queen safely in a cage, you can now install her following the procedure outlined below.

If you are harvesting your own swarm cells, you may notice that a mature cell has bees crawling all over it, shaking the cell, while also chewing the wax off the tip. The shaking behavior is referred to as vibratory signaling, where the workers are communicating with the queen, and is a sure indication that the queen inside is alive and well.

Removing your own queen cells from comb requires a gentle touch. Queens, like all capped pupae are fragile, and the cell must not be deformed in the process of removal. It is best removed down to the midrib of the comb, with enough wax so that no royal jelly is exposed underneath. They must also be kept warm and at the same time not overheated while in transition to their new colony.

CHAPTER FOUR

VIRGINS AND THE PARENT COLONY QUEEN

Finding virgin queens for sale may be a challenge, but, in terms of timing their introduction, they are the most like swarm cells and walkaway splits. The technique is the same, meaning you must delay the introduction long enough to facilitate the same length brood break. Just like with ripe queen cells, there are the same critical timing issues and colony behaviors to consider. If the split had eggs, the colony will start emergency cells almost immediately. Those will need to be removed as explained previously. If no emergency cells are present, you can simply leave the split queenless for 6 days, then introduce the new virgin queen in her cage with the cork in place denying access to the candy plug normally used to release her. After about 6 days you can remove the cork and expose the candy plug. Then it takes another 3-4 days until she's released. After the virgin queen is released, she will follow the same mating events as a walkaway or ripe queen cell split, meaning she will take orientation flights and when conditions are favorable, take her mating flight.

Dealing with the Parent Colony Queen

In dealing with the parent colony queen, there are a few things to consider. If you're splitting because you've detected high varroa populations and you suspect the queen is producing workers with low hygienic behavior, eliminating her from your apiary may be the best option. On the other hand, you may be splitting as part of routine IPM practices, and the queen has qualities that you want. In that case, there are a few options for handling her.

If you've made multiple splits from a single colony and don't have an immediate need for the parent queen, you may want to put her in a very small nuc box with just one frame of open brood, some nurse bees, a food frame, and fill the remainder of the box with drawn comb or foundation. That way, she can continue to lay, stay healthy, and be available if one of your other re-queening options fails.

If you're making a 50/50 split, taking one small split from a colony and want varroa control for the larger part with the parent queen, or splits are not desired or possible, then you will need to restrict the queen from laying. If your goal is an extended brood break that simulates the other re-queening procedures, then, as mentioned previously, the colony must go eggless for about 27 days.

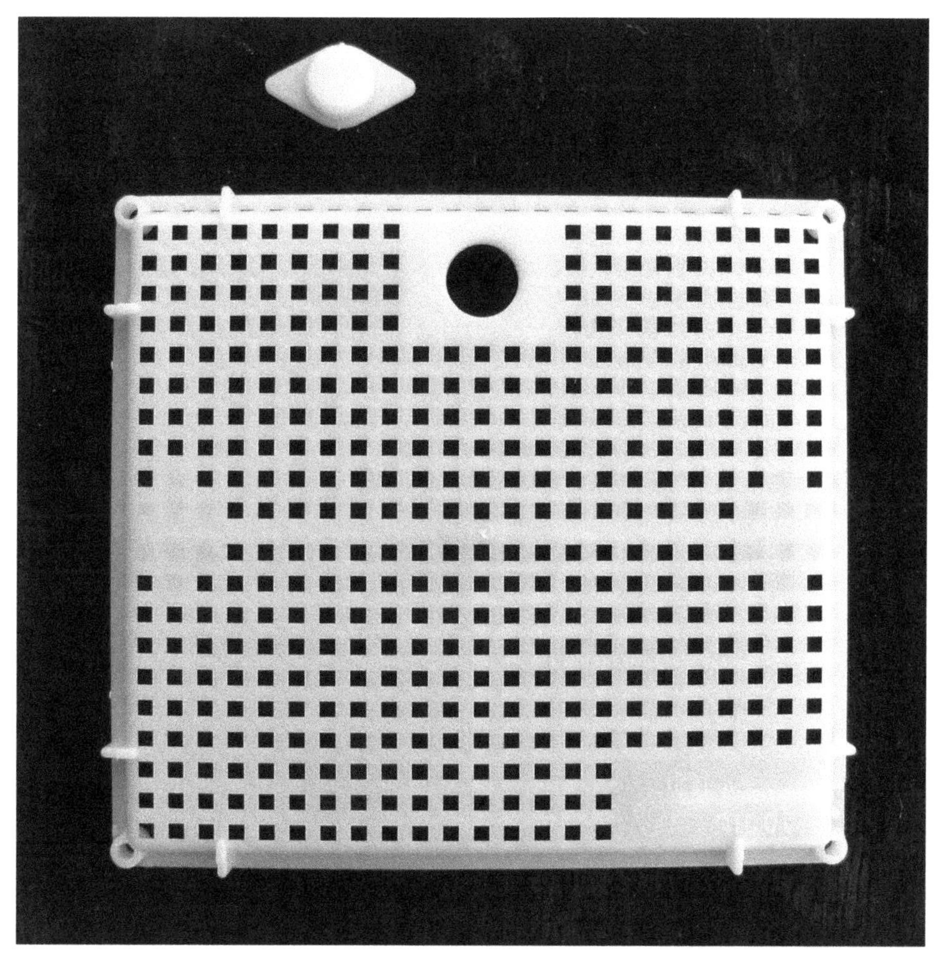

Figure 5: *A commercial push-in cage like this one will allow the existing queen to remain in the colony, but restrict her from laying eggs beyond the boundaries of the cage. They are also useful when introducing a new queen to a colony.*

Figure 6: *When making your own push-in cage from #8 hardware cloth, it must be deep enough to reach the mid-rib of the comb and, at the same time, allow room for the queen to move freely underneath.*

The healthiest procedure is to leave the queen in the colony, but restrict her laying to a small area under a "push-in cage" (Figure 5). These cages are commercially available or can be made from #8 hardware cloth (Figure 6). When using a push-in cage, a comb from the colony that has some capped brood and open cells is selected. After all the adult bees are brushed from the comb, the cage is pushed in over a small amount of capped brood and open cells, making sure none of the colony's adult bees are under the cage. The queen is then placed in the cage, and the emerging bees care for her until she is eventually released.

Her physical presence in the colony will maintain order, and you can leave her caged until all the existing brood has emerged. Since the colony thinks it's queen-right, they should not make emergency cells, but it's a good practice to go back in and check a few days later.

CHAPTER FIVE

CHOOSING OPTIONS AND OTHER FACTORS

To help decide, start with an inspection and determine the number of splits you can make, based on the quantity of nurse bees, brood, and food frames that are available in that colony. Choosing an option, like queen cells or virgin queens, will depend on connecting with a supplier and timing the splits with delivery. During swarm season you can use your own swarm cells, and, as mentioned, a walkaway can be done almost spontaneously, as long as drones are available to mate with the virgin queen.

All the options require inspections to determine if the queen is accepted and laying. When re-queening doesn't happen according to plan, you must intervene to get the colony on a path to becoming queen-right. Leaving a colony without a queen for an extended period will result in workers developing their ovaries, and when that happens, they will start laying eggs (Figure 7). Since worker bees are not fertile, a colony of laying workers produces all drones, and the colony is basically doomed without intensive manipulations to encourage them to raise a new queen or accept a purchased queen. Also, if a virgin queen from any option is not mated correctly she can also lay all drones and must be replaced.

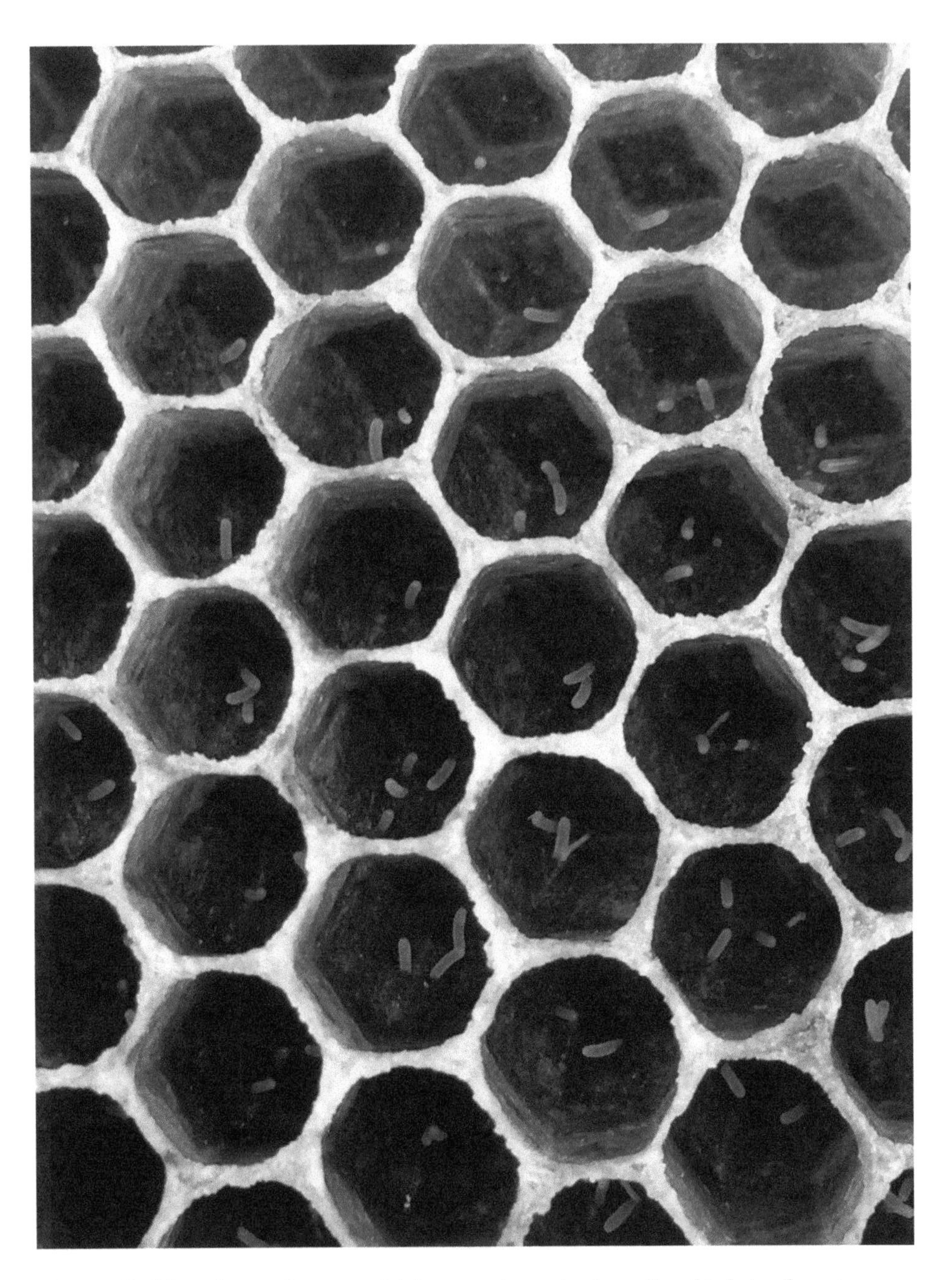

Figure 7: *This photo shows multiple eggs in a single cell, which is almost a sure sign that the colony has developed laying-workers. At this point the colony has been queen-less too long and will be difficult save.*

Weather

The weather plays a major role in the mating of virgin queens and can also delay your planned date for doing splits. Queens will emerge from cells on their natural schedule regardless of weather, so you may need to think about how you would care for virgin queens in case weather prevents you from getting cells into colonies before they emerge. Keep notes on the days when your new queens couldn't fly, and adjust your timing accordingly. A rainy day is of little consequence, but pay attention to long rainy periods or unseasonably cold weather.

Swarms as Brood Breaks

Swarms and walkaway splits are similar, with the exception that swarms are planned reproductive events completely managed by the colony. They occur in a specific way, attempting to ensure the success of both the swarm and the parent colony. With reproductive swarms, queen cells are raised about 8-10 days before the swarm occurs, which minimizes the queenless period. Also, at some point prior to the swarm, the existing queen is often restricted from laying and thinned down, so she can fly and leave with the swarm. If the colony has a single swarm event and no more, a new virgin queen will emerge, mate, and start laying in about three weeks. That's a few days earlier than the approximately 27-day interval discussed earlier with a walkaway split. This is because the swarm has the advantage of advanced queen cell development, while a walkaway is developing a queen from an egg, or a day old larva, beginning on the first day of the split. That gives the swarm a head start in queen rearing, although their actual broodless period will be similar because, as mentioned, the queen that exits with the swarm has usually been restricted from laying prior to the swarm event.

Keep in mind that during reproductive swarms the colony is balancing the bee population in terms of the quantity of capped brood and determining if an after- swarm will occur. It's not unusual for a strong colony, with an abundance of capped brood and bees, to issue one or more after-swarms. When after-swarms do occur, the queen-less interval is extended, while one or more virgin queens emerge and the last swarm leaves. This can result in a very long queenless period and, therefore, an extended broodless period. It's important to note that the same thing can occur with an over populated walkaway split done during swarm season. Swarms are common even after a 50/50 walkaway, when a beekeeper basically splits a strong colony in half and places an abundant amount of capped brood and bees in both. One theory is that a populous split is advanced into simulated swarm behavior and, with an abundance of capped brood, will swarm with the first virgin that emerges and may continue to swarm until the colony achieves a natural balance. Therefore, if you want to give your splits the best chance at avoiding swarm behavior, avoid a large 50/50 split with lots of capped brood and make a few smaller ones.

NOTES

Since these splits are for the purpose of varroa control, monitoring mite populations before and after is critical to understanding efficacy. If your ultimate goal is limited treatment, your re-queening choice should favor queens with hygienic behavior that resist varroa or, conversely, bees that you have learned can tolerate higher levels of infestation. For reasons still unknown, some bees will develop a measurable mite resistance over time in a given location and, unfortunately, only in that location. When moved, they seem to lose their resistance, so be aware that this possibility may occur for a colony in your apiary and those queens will be quite valuable.

When using walkaway splits, ripe queen cells, or virgin queens, area drones will determine part of the genes passed on to the splits. Therefore, the hygienic quality of locally mated queens can only be determined as you observe the varroa counts in those colonies. Also, you will lose queens purchased for their hygienic traits if they swarm, so record and note swarm dates, because those colonies will require that you evaluate the new daughter queens.

Splits for varroa come at a cost to both the colony's strength and the ability to make honey. There's an old saying "you can make honey or make queens, but not both." It's a little simplistic, but the point to remember is that splits set a colony back. Even so, if you time your splits so they are done early in the season, your colonies will recover in plenty of time. Healthy bees can overcome lots of obstacles.

Splits also increase the size of your apiary, and that may not be desirable. In that case, your option is to recombine the splits into larger colonies later in the season. You can manage the growth by removing resources while the split grows. There's a lot of opportunity to remove brood, or you can use the queen to re-queen another colony and let them raise a new queen, or harvest queen cells.

Using splits to control varroa is a powerful natural tool, but unfortunately with varroa, no single pest strategy is enough. I encourage you to monitor your mite counts and use splits as part of a larger IPM program designed to keep your bees healthy. Until our European honey bee strikes a natural balance with varroa, beekeepers are in the role of animal husbandry and whether you view bees as domestic stock, pets, or precious global pollinators, we have an obligation to care for them. Good luck with your bees!

APPENDIX ONE

VARROA POPULATION GROWTH DYNAMICS

Varroa grow exponentially, meaning that not only does the population increase with time, but the percentage of population growth increases also- here's how it happens. An adult female mite will enter a cell when the honey bee larva is hours from being capped. The cell is then capped, and the female, or foundress mite, begins to lay eggs. The first egg is a male, and then at 30-hour intervals she lays female eggs. As the females emerge, the male mates with his sisters, and if the eggs were laid in a worker cell, about 1.5 female mites mature and emerge with the bee. Since a drone takes approximately 3 days longer to emerge, mite eggs laid the same way in a drone cell can produce about 2.2 mature females. As a result, when drone cells are available, varroa are about 8 times more likely to select them for reproduction.

To put the numbers in practical terms, you can figure that in 12 weeks, not considering swarms and splits, the number of mites can multiply by roughly 12. Therefore, starting with a population of 20 mites, the population can grow to approximately 240 in the first 12 weeks. At first glance that's not so impressive, but exponential growth is deceptive, because the real population explosion happens disproportionately with time. Twelve weeks later, 240 mites can grow to approximately 2880. This rarely happens in the first year of a colony's life, but will happen in time if the colony is left untreated. Once the mite population passes 2000, things really heat up, and the mite population can easily approach 5% -8% of the colony's population of bees. With the current recommendation to keep levels at or below 2%, colonies approaching double that amount can begin to exhibit symptoms of Parasitic Mite Syndrome (PMS). Full PMS means the bees are no longer able to manage the viral load vectored by the mites. The result is bees with deformed wings, brood unable to mature, and a number of other maladies that lead to the death of the colony.

APPENDIX TWO

CRITICAL DATES

Critical Dates Post Split- Approximate

Day 9	All existing open brood are capped
Day 12	New queens emerge from various options
Day 24	All brood and all mites have emerged
Day 24	All mites are phoretic until 7-8 days after the new queen starts laying

Critical Dates Based on the New Queen – Plus or Minus 5 days

Day 27	New queen starts laying
Day 33	Last day that all mites are phoretic
Day 34	Mites resume reproduction in new brood
Day 48	New mites begin to emerge

REFERENCES

Evans, Kathleen. "*Evaluation of early summer splits on varroa mite reduction and colony productivity.*" Diss. University of Delaware, 2015.

Graham, Joe M. "*The Hive and the Honey Bee.*" Dadant & Sons, 2015.

Jacobson, Stu. *"Locally adapted, varroa resistant honey bees: ideas from several key studies"* Am Bee J 150 (2010): 777-781.

Oliver, Randy. *"Sick Bees – Part 12: Varroa Management – Getting Down to Brass Tacks"* scientificbeekeeping.com. 1 Sept. 2011. Web. 24 June. 2014.

Pérez-Sato, Juan Antonio, and Francis LW Ratnieks. *"Comparing alternative methods of introducing virgin queens (Apis mellifera) into mating nucleus hives."* Apidologie 37.5 (2006): 571-576.

Sammataro, Diana, Uri Gerson, and Glen Needham. *"Parasitic mites of honey bees: life history, implications, and impact."* Annual review of entomology 45.1 (2000): 519-548.

Sammataro, Diana, and Alphonse Avitabile. *"The beekeeper's handbook."* Cornell University Press, 1998.

Schneider, S. S., S. Painter-Kurt, and G. DeGrandi-Hoffman. *"The role of the vibration signal during queen competition in colonies of the honeybee, Apis mellifera."* Animal behaviour 61.6 (2001): 1173-1180.

Winston, Mark L. *"The Biology of the Honey Bee."* Harvard University Press, 1991.

CPSIA information can be obtained
at www.ICGtesting.com
Printed in the USA
BVHW02s2100040318
509336BV00013B/100/P

9 781908 904867